# MEASUREMENT THEORY

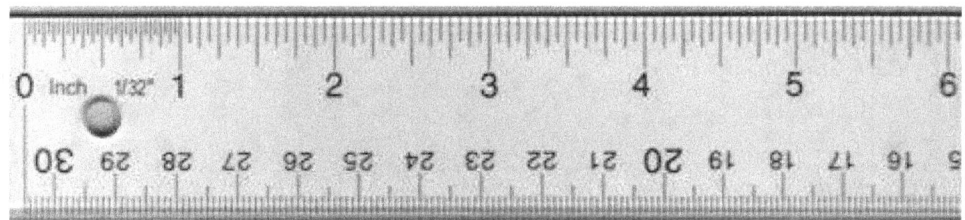

## MATHEMATICS MADE EASY: MEASUREMENT THEORY

Authored by Mrs. Paula Burrows

Published by BSM Consulting.

Copyright © 2015.

All Rights Reserved.

ISBN:1511456914

For orders and inquiries, please send to:

BSM Consulting
P.O. Box EE15057
Nassau, Bahamas
Email: bsmlifeconsult@gmail.com

## ABOUT THIS BOOK

This book is written for Secondary Mathematics students with challenges. Many students find the subject challenging. Mathematics Made Easy is designed to build students' confidence in solving Math problems. This book was written by Paula Burrows and inspired by her time in the classroom teaching Mathematics at the secondary level.

***Mathematics Made Easy: Measurement Theory*** is the 5th book in the '*Mathematics Made Easy*' Series. Other books in this series include:

- Mathematics Made Easy: Number Theory (Book 1)
- Mathematics Made Easy: Number Theory Answer Key (Book 2)
- Mathematics Made Easy: Geometry Basics (Book 3)
- Mathematics Made Easy: Geometry Basics Answer Key (Book 4)
- Mathematics Made Easy: Measurement Theory Answer Key (Book 6)
- Mathematics Made Easy: Mensuration (Book 7)
- Mathematics Made Easy: Mensuration Answer Key (Book 8)
- Mathematics Made Easy: Algebra (Book 9)
- Mathematics Made Easy: Algebra Answer Key (Book 10)
- Mathematics Made Easy: Number Sets (Book 11)
- Mathematics Made Easy: Number Sets Answer Key (Book 12)
- Mathematics Made Easy: Symmetry and Graphs (Book 13)
- Mathematics Made Easy: Symmetry and Graphs Answer Key (Book 14)
- Mathematics Made Easy: Fractions (Book 15)
- Mathematics Made Easy: Fractions Answer Key (Book 16)
- Mathematics Made Easy: Decimals (Book 17)
- Mathematics Made Easy: Decimals Answer Key (Book 18)
- Mathematics Made Easy: Percentages (Book 19)
- Mathematics Made Easy: Percentages Answer Key (Book 20)
- Mathematics Made Easy: Ratio and Proportions (Book 21)
- Mathematics Made Easy: Ratio and Proportions Answer Key (Book 22)
- Mathematics Made Easy: Probability and Statistics (Book 23)
- Mathematics Made Easy: Probability and Statistics Answer Key (Book 24)

For a complete book containing all topics, see Practicing Mathematics for BJC Success. It contains practice questions on all the topics in the Mathematics Made Easy Series.

# TABLE OF CONTENTS

## PART I - CORE QUESTIONS

# DRAW AND MEASURE LINE SEGMENTS OF DIFFERENT UNITS

**EXERCISE 1**

1.  Draw line segments with the following measures:

(a)  4cm                                          (b) 3 inches

(c)  65mm                                         (d)  2 ½ inches

2.  Measure and write down the length of AB in millimeters.

A _____B

Answer: _____

3.

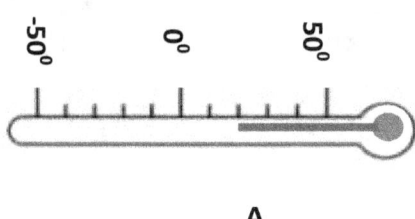

**A**

(a)  Write down the temperature shown at **A**.


Answer:  _____


(b)  Mark and label the spot **Y** which is $25^0$ below 0.


4.                                                                                              NOT TO SCALE

(a)  How many inches does the line segment **IP** represent?


Answer: _____

(b)  Which letter represents 2 ½ units from **J**?

Answer: _____

5.  (a) Measure the span AB to the nearest millimeter.  (BJC2001 p.2/6)

(b) If 1mm represents 5cm, write down the length of the span.

Answer: _____

# CONVERT METRIC UNITS

To help you remember the order of metric prefixes remember the following phrase:

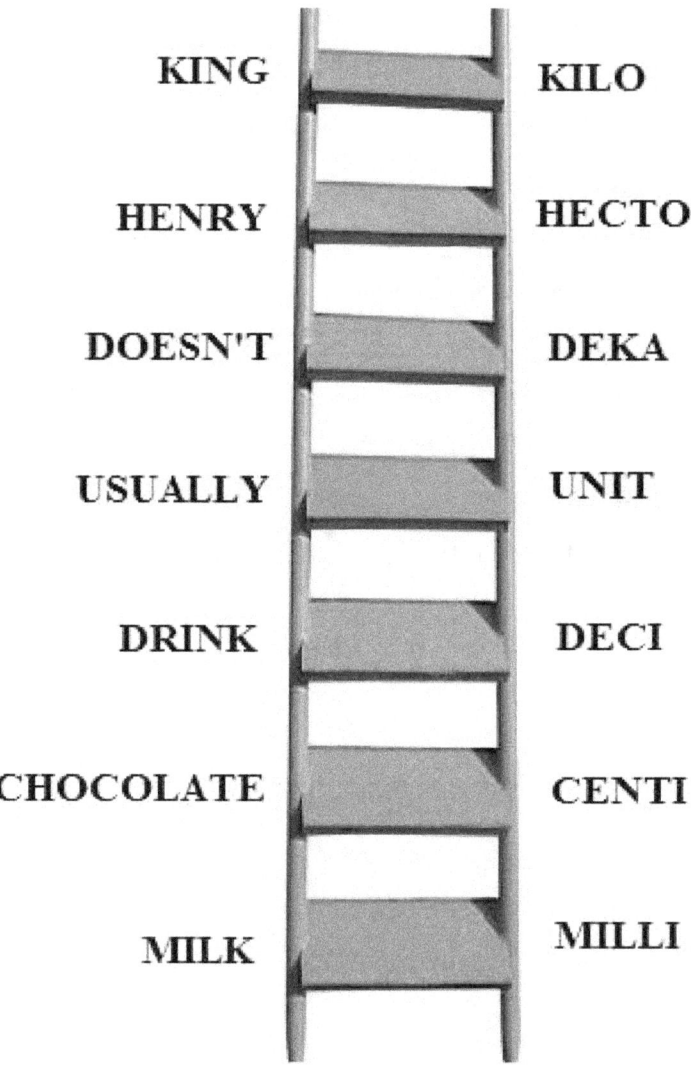

| | |
|---|---|
| KING | KILO |
| HENRY | HECTO |
| DOESN'T | DEKA |
| USUALLY | UNIT |
| DRINK | DECI |
| CHOCOLATE | CENTI |
| MILK | MILLI |

**EXERCISE 1**

Convert each measurement to the unit in brackets.

(a) 50mm (cm)

Answer_____

(b) 460cm (m)

Answer_____

(c) 3.5km (m)

Answer_____

(d) 5.67m (mm)

Answer_____

(e) 4000m (km)

Answer_____

(f) 8.9m (cm)

Answer_____

(g) 50m (km)

Answer_____

(h)  3km (mm)


Answer_____


## EXERCISE 2

1. One watermelon weighs 1.3 **kg**. Another watermelon weighs 110.4**g**. What is the

total weight, in **grams**, of both watermelons?


Answer _____


2.  Write down the lengths in descending order.

**7 cm, 20 mm, 3m, 5 dm**


Answer: _____

# READ/ANALYZE TIME USING CLOCKS AND/OR CALENDARS

**EXERCISE 1**

1. Study the calendar below, and then answer the questions that follow:

NOVEMBER

| S | M | T | W | Th | F | S |
|---|---|---|---|----|---|---|
|   |   |   | 1 | 2  | 3 | 4 |
| 5 | 6 | 7 | 8 | 9  | 10 | 11 |
| 12 | 13 | 14 | 15 | 16 | 17 | 18 |
| 19 | 20 | 21 | 22 | 23 | 24 | 25 |
| 26 | 27 | 28 | 29 | 30 |   |   |

(a) Andre's birthday is on the third Tuesday in November.

What date would that be?

Answer: _____

(b) What day of the week is two weeks after the 15$^{th}$?

Answer: _____

(c) The Sunday nearest to November 10$^{th}$ is Remembrance Sunday. What date was Remembrance Sunday?

Answer: _____

## PART II - EXTENDED QUESTIONS

# CALCULATE TIME ELAPSED

**EXERCISE 1**

1. A cooking lesson that lasted for 2 hours and 15 minutes ended at 2:05pm. At what time did the lesson begin?


   Answer: _____


2. In 1983, Mrs. Bethel was 56 years old. In what year was Mrs. Bethel born?


   Answer: _____


3. Bahamasair leaves Nassau at 7:10 am and arrives in Mayaguana at 9:05am. How long does the flight last?


   Answer: _____

4.  Job Starts          Job Ends

(BJC2004 p.1/19)

(a) At what time did the job start?


Answer: _____


(b) What time did the job end?


Answer: _____


(c) How long did the job last?


Answer: _____

5.  Lunch on board the SS Ocean Link is served at 1500 hours.  Express this time using the 12 hour clock.

Answer: _____

6.  How many days are there from the 14th August 12:00 noon to the 22nd October, 12:00 noon?

Answer: _____

# COMPLETE SCALE /ACCURATE DRAWINGS

**EXERCISE 1**

1.

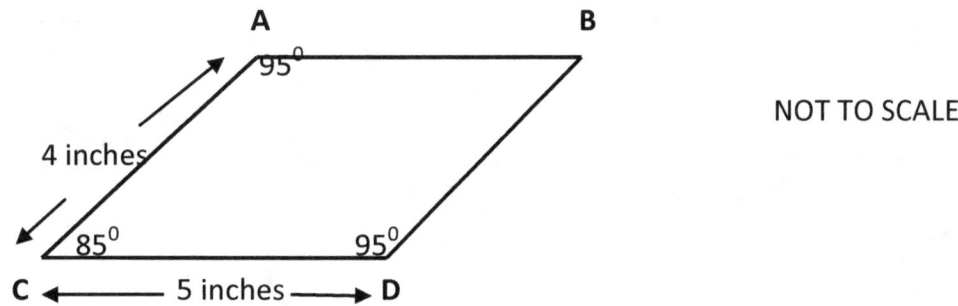

NOT TO SCALE

(a) Using the above sketch, make an ACCURATE drawing using the measurement theory on the sketch.

(b) Give the special name of this quadrilateral.

Answer_____

2.   **LMNO** is a rectangle.

(a) Draw in a diagonal.

(b) Measure the length of the diagonal in centimeters.

Answer_____

(c) Bisect the diagonal, using your compass and pencil.

A circle is to be drawn around the rectangle

(d) State the length of the radius to be used.

Answer_____

(e) Draw the circle.

3.

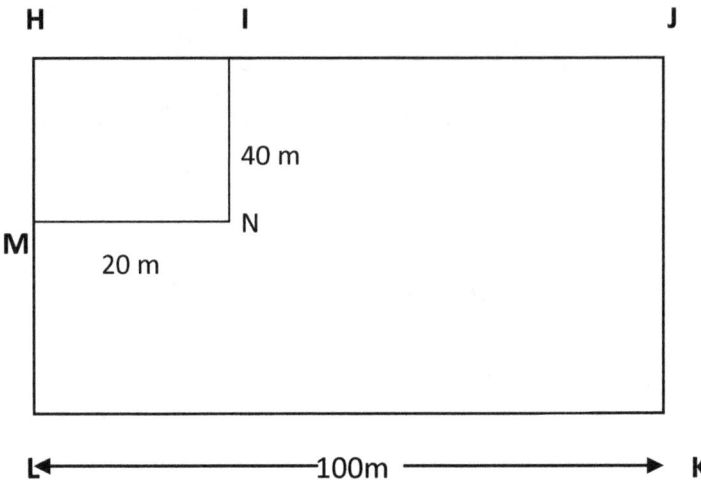

The figure is a plan of a school gymnasium with a coach's office in one corner.

(a) Draw the plan **to scale** using 1 centimeter to represent 10 meters.

(b) Write down the actual distance of **KN**.

Answer _____

4.

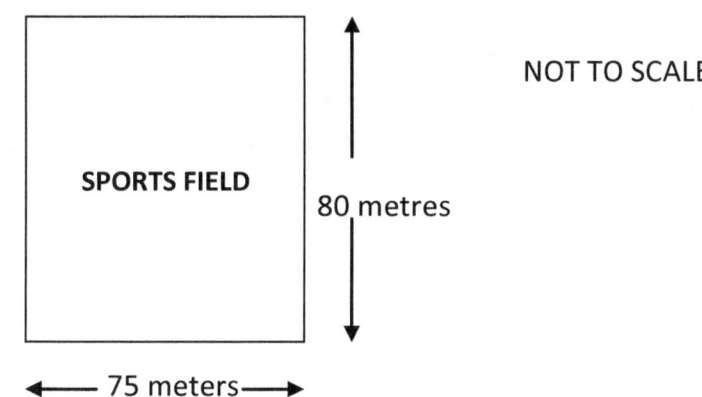

NOT TO SCALE

(a) Using a scale of 1 cm to 10 m, make an accurate drawing of the sports field above.

(b) Draw in a diagonal in your rectangle.

(c) Measure then write down the actual length of the diagonal.

Answer _____

# PART III – ANSWER KEY

## ANSWER KEY

**DRAW AND MEASURE LINE SEGEMENTS OF DIFFERENT UNITS**

**Exercise 1**

1. (a) Draw image based on your rule measurement
   (b) Draw image based on your rule measurement
   (c) Draw image based on your rule measurement
   (d) Draw image based on your rule measurement

2. 80mm (though measurement can vary depending on source document)

3. (a) 3.5 inches

   (b)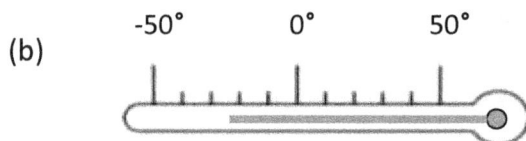

4. (a) 3.5 inches

   (b) The letter 'O'

5. (a) 50mm (or based on your rule measure)

   (b) 250cm

## CONVERT METRIC UNITS

### Exercise 1

(a) 5 cm

(b) 4.6 m

(c) 3,500 m

(d) 5,670 mm

(e) 4 km

(f) 890 cm

(g) 0.05 km

(h) 3,000,000 mm

### Exercise 2

1. 1410.4 g

2. 3m, 5dm, 7cm, 20mm

For the complete answer key for this series, please reference the book below:

**MATHEMATICS MADE EASY: MEASUREMENT THEORY ANSWER KEY (BOOK 6)**